AF863754

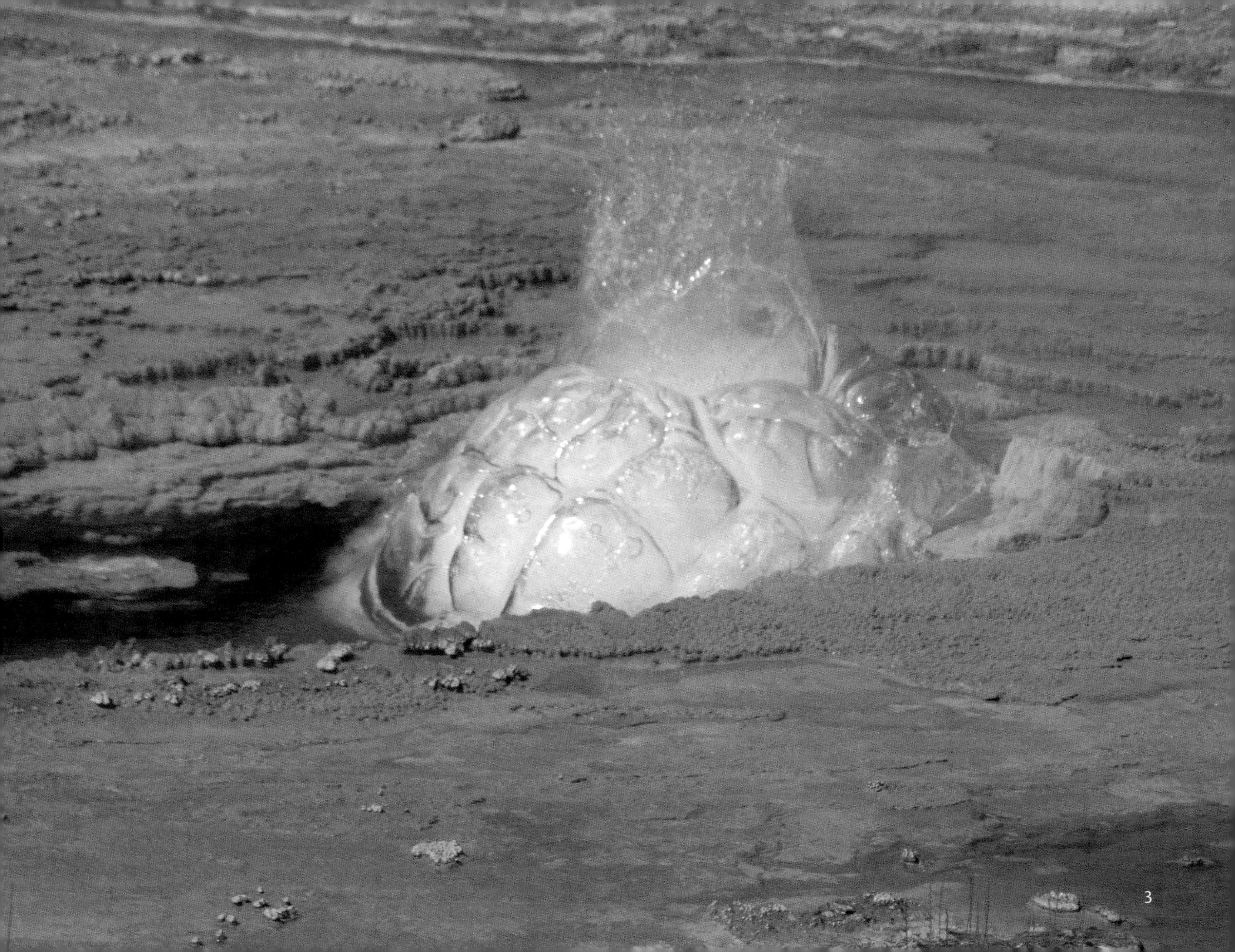

Closeup of a Botryoidal Spring bubble beginning to break up, showing surface texture that appears glass-like. Soon it will be a fountain of spray like all other geysers.
Previous pages show a sequential time history of a Botryoidal eruption from a nearby hilltop. Pictures cover 1 second of the eruption. After it turns into a mound of bubbles, a cylinder of spray rises up shaped like a volcanic eruption. In the final image bubbly spray is rising in the cylinder and the mound of bubbles has turned into a ring of giant bubbles, resulting in an unusual and beautiful image.

Iceland's Strokkur Geyser starts with a blue bubble propelled by rising steam bubbles. An area of superheated water is visible to the left of the cauliflower-shaped central bubble, and water is flowing out with a velocity sufficient to produce small areas of cavitation bubbles hugging the crater lip. These are produced by the rapid outflow of water from the rising dome. In this picture, all features of a blue bubble are visible at once. Textured geyserite protrudes from the runoff channel.

Strokkur's water column has broken free of its bubble and is rising skyward. Soon it could reach 100 feet (30 m) or more.

Its the end of May, yet the volcano in the background has snow extending nearly to sea level and the road north across the center of Iceland is still impassable.

Unusual Geysers,
Hidden Patterns in Geysers of Yellowstone & Iceland

Copyright © 2016 by Jack Hobart, in softbound and e-Book (pdf) formats.
All photographs by the author, unless noted.

ISBN 978-1-944564-01-8
Printed in the U.S.A. by Sheridan Press

<u>Front Cover</u>:

Unusual Geysers

Hidden Bubble Patterns in Geysers of Yellowstone & Iceland

Image: Botryoidal Spring in the Lower Geyser Basin, White Creek Group near Great Fountain Geyser.
A blue bubble pattern is shown at its peak before turning to spray. The pattern seems to be covered a with heart-like texture all over its surface.

Note: geyser pictures in the all but the final chapter are of Botryoidal Spring unless noted.

<u>Back Cover</u>:

Unusual Geysers

Strokkur Geyser, Iceland's Crown Jewel

Images: Strokkur Geyser at Haukadalur near the center of the country. It is adjacent to the Great Geysir for which all geysers are named.
Shown are a blue bubble for which it is famous, a full eruption that is rising to over 30 m (100 ft) in height and a slow camera shutter speed eruption that illustrates the portions of the burst which are expanding fastest from its blue bubble. That pattern is reminiscent of a fleur-de-lis, the European coat of arms most associated with the French monarchy for which it means flower of the lily.
Note: all pictures in the final chapter are of Strokkur, unless noted.

At right: Strokkur's burst at full height can be so beautiful that it deserves a custom frame

Unusual Geysers

Hidden Bubble Patterns in Geysers of Yellowstone & Iceland

Jack Hobart

Enjoy!! ☺ Jack Hobart

This closeup view of Botryoidal Spring shows the complex pattern and texture that cannot be seen in geysers having normal spray eruptions. Only a few percent of all geysers ever exhibit bubble patterns long enough to be visible to a casual observer.

contents

from the author

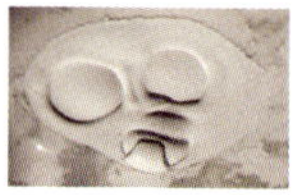

As a child of 9, I was brought to Yellowstone by my aunts. That experience started a lifelong fascination with the Park and its ever-changing showcase of thermal features. For more than 30 years, I have been taking photos and videos of its mudpots and geysers with my sons Craig and Chad, plus extensive help from many friends who likewise share a love of Yellowstone.

Most of these people are members of GOSA, the Geyser Observation and Study Association, an organization with worldwide membership. It publishes the Sput newsletter plus the GOSA Transactions, a journal of scientific and descriptive articles, published every 2 or 3 years. They also offer reprints of many related publications by others including monumental compilations that can exceed 1000 pages.

The study and enjoyment of these geothermal features is infectious, leading to many friendships and a comradeship that helps learn much more than possible from simply reading a guidebook or visitor's brochure. These people plus all the geyser gazers of GOSA constitute the primary resources for maximizing the experience of these geothermal treasures. Hope you also run into them in the Park.

A number of people deserve special mention for their extraordinary contributions over the years.

Rocco Paperiello and the late Marie Wolf compiled the most exhaustive descriptions of Yellowstone's thermal features including the 1,500 page "Lesser-Known Thermal Units of Yellowstone National Park". He is also an encyclopedic compendium of descriptive and historical information.

Lynn Stephens is an equally bountiful source of current and past information about geyser characteristics.

T. Scott Bryan edited the GOSA newsletter for the first decade and publishes The Geysers of Yellowstone, the go-to resource describing all geysers in Yellowstone plus descriptions of the World's geyser regions.

Tom and Genean Dunn took over publication of the Sput in 1996 and have tirelessly brought it up to the classy full color paper and electronic newsletter of today.

The Park Chief Geologist, the late Rick Hutchinson, was a supporter of the study of obscure park geothermal features including the exploration of backcountry explosion craters. The Park Service and its rangers also provided invaluable assistance through the years, including permission to study Botryoidal Spring from poolside on a number of occasions.

The Goldbergs, Roz, Charlie, Dave and Mike also explored many of these features with me. We enjoyed many special moments. Miss you Charlie.

Mike and Cynthia Keller have been close friends and advisors for most of my Yellowstone years as well. Their daughters Alex and Megan are particularly adept at discovering hidden faces and features within mudpot explosion and bubble geyser eruption photographs such as these.

Purpose of this book. My goal is to show images of eruption starts of those rare geysers that produce initial bubble domes that soon turn into kaleidoscopic arrays of patterns and textures. These can only be captured with very high speed photography. Many don't seem real. They appear to be art masterpieces chiseled from rock crystal rather than simply made of liquid water.

Lets go exploring !

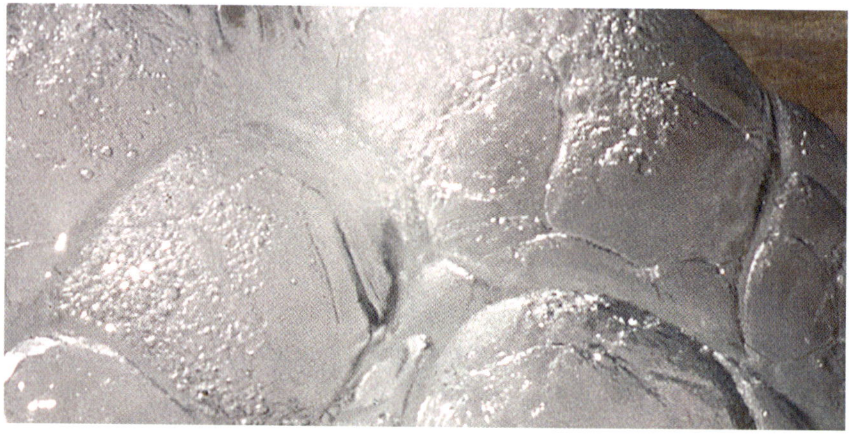

what is a bubble geyser?

Geysers are hot springs that intermittently spray water above their surfaces. These water surfaces can be above or rarely, below ground. The force behind their operation is superheated water whose temperature rises due to heat flow from a body of volcanic heat at great depth.

Eventually steam bubbles form and rise toward the surface, accelerating as they rise, usually causing a rise in the level of the geyser's water surface and possibly causing an overflow from its pool or crater.

If conditions are right, the bubbles cause the water to dome up before they break the surface. This dome is commonly referred to as a blue bubble, named after the beautiful bubbles seen in a few geysers. Probably only a few percent of all geysers produce bubbles that last long enough to be easily observed by those waiting for a normal eruption from a particular geyser. Below is an exceptional example from Strokkur Geyser in Iceland. It produces blue bubbles on a regular basis. Many more will be shown in the chapter on Strokkur.

The top of this rising bubble is just becoming visible above the crater lip, and water is uniformly overflowing on all sides.

Some geysers erupt from a relatively broad pool, so one or more bubbles are able to rise at different locations, sometimes simultaneously. Botryoidal Spring in Yellowstone is the premier example of such a pool. The resultant variety in its bubbles makes it the most exceptional bubble geyser in the world - and my favorite by a wide margin. Most of the pictures in this book are from this geyser. I'm on a first-name basis with it so I'll usually refer to it that way.

Some geysers have only a small number of eruptions in which water doming is visible before a burst. Generally, it only occurs in the more forceful eruptions. These may be accompanied by an audible "pop" just before or as the dome breaks apart. Because this is so unexpected, few pictures have ever been taken of blue bubbles from these geysers. The advent of high quality cell phone video makes it much easier to capture these blue bubbles, so they are appearing more regularly on You Tube and in the Sput.

The best examples of such geysers are Great Fountain, Fountain and Morning geysers. Their audible "pops" usually indicate that a blue bubble has occurred. Photography usually shows a beautiful semicircular spray from what was a subsurface or surface blue bubble. These pops are probably the result of water slamming shut on an orifice created as a steam bubble escapes from a water surface. I have photographed these geysers hundreds of times and never been able to capture a single blue bubble.

I had a memorable experience at one geyser that is noted for its remarkable blue bubbles. It is called Deep Blue and is located at the far side of the Kaleidoscope Group of Lower Geyser Basin. Huge blue bubbles have been reported, associated with 40 foot eruptions, possibly much higher.

One year, the Park Service allowed knowledgeable observers to go to that side of the basin, however you had to approach from the Firehole River, a mile long trek from the nearest road. Access was only allowed in the early morning, when mist from the many thermal features totally enveloped the area. When fog began to lift, you had to vacate the area before there was any chance of being seen from the main road or a boardwalk.

It was quite spooky standing some distance from the geyser, waiting for the next eruption. I had never seen it before, so wasn't sure what to expect. Suddenly, directly below me, a loud pop occurred, I was startled to say the least, however the following eruption was not accompanied by a blue bubble. Soon, the fog began to lift, and I lost my only chance to see a bubble from this geyser.

Little has been written about eruption bubbles, so I've been forced to develop my own terminology to describe them. I define the bubble as an unbroken water surface, visible above the water level or crater lip of a geyser. This surface can be quite irregular, including intersections of smaller bubbles as you often see when blowing soap bubbles.

Eventually the bubble breaks up when the internal gas escapes from the water surface. I refer to this as occurrence of cavitation at the bubble surface. This description of water and gas leaving the surface may not be explicitly correct from a fluid dynamics point of view, but it enables me to define the time at which each portion of the bubble surface breaks up. A number of examples are shown below.

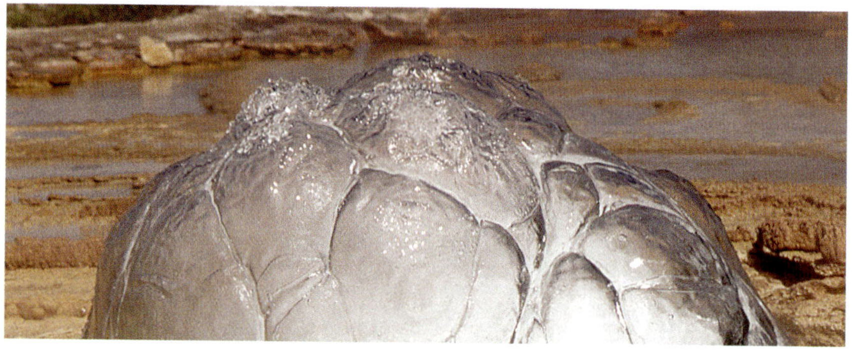

Many early eruption sequences are shown in the book. These give a good idea of what constitutes the beginning and the end of the blue bubble phase. I refer to them all as blue bubbles, even if some are white due to their internal steam content.

An unusually large bubble with rear half highlighted against the shadows behind it. Water has risen sufficiently to overflow the rim.

Following two pages: a sequence showing how 5 separate rising bubbles coalesce (merge together) as they push above the water surface. Notice how much the water level rises between the two pictures.

Water flow patterns are clearly visible. It is being pushed up against the far side enough to produce splashing. It is also flowing away from the mass of bubbles rising up across a wide area

An example of a very large, 3-part bubble that extends over the sides of the pool without beginning to burst forth into its eruption spray.

Closeup of a blue bubble showing deep furrows all over. The dome of water retains ripples from the surface of the pool.

discovering geysers that bubble

For decades, I have been known as the mudpot guy in Yellowstone because I spent all my time taking pictures and video of exploding mudpots and mud geysers. This requires waiting patiently for the right moment, then taking photographs at very high shutter speeds to capture the motion. Many videos were also taken at high shutter speed, so that the expanding mud could be studied one frame at a time.

The following examples indicate similarities between pictures of mudpot explosions and pictures of bubble geyser eruptions.

A mud explosion in a backcountry hydrothermal explosion crater, a 50 foot diameter mudpot that was formerly a grove of trees. Mud was flying up to 100 feet high.

A steam explosion propels mud from a vent much like the change from a geyser bubble to a spray eruption. 'Pops" in mud are far more common than similar noises produced by bubble geysers.

This is actually a mud geyser burst, creating a beautiful pattern that dissipates in a fraction of a second, just like a geyser blue bubble

You can tell from these pictures that I focus on shapes and patterns, whether they be in mudpots, geysers or actually most anything that I come across.

The mud version of a geyser bubble eruption.

Prior to 1996, geyser photography was not much of an attraction to me. I was attracted to especially large bursts, eruptions of rare geysers, dramatic lighting and capturing special moments with family and friends. All that was about to change - with important consequences for my priorities .

Geyser observation and monitoring prior to 10 years ago was quite different than today. You would take some pictures, return home, then see if you got any good pictures after a week or so. This discouraged people from spending lots of time with experimental photography.

There were stories of a few instances in which geysers would dome up, then explode to an exceptional height, but none were known to reliably produce what is now known as a blue bubble. Even an article about Iceland geysers in the 2002 GOSA Transactions failed to mention bubbles from Strokkur.

That changed when Botryoidal Spring began erupting as a geyser. Previously, it was a boiling hot spring, 100 yards southeast of the Fountain Lake loop road at Great Fountain Geyser. It was part of the White Creek group of geothermal features. The extremely hot White Creek flows past to its south and the former White Creek-Midway Geyser trail passed by to its north, just below a bluff marking the edge of the geyser basin.

Botryoidal Spring was a perpetual spouter, bubbling constantly and rarely boiling up to a height of as much as 7 feet, therefore it was classed as a perpetual spouter and sometime geyser in early versions of Scott Bryan's Geysers of Yellowstone book. This boiling produced grape-like geyserite nodules around its rim, hence the name Botryoidal.

In 1996, Botryoidal activated and became a regular, periodic geyser. At the time, an earthquake was said to be the cause, however no significant earthquakes occurred nearby.

My older son Craig and I investigated and found the geyser erupting every 1.5-2 minutes, suddenly bursting up to as much as 15 feet, gradually tailing off to quiet within 15 seconds. Some of the eruptions started with a doming of water that turned to a dome of bubbles..

This intrigued me, since they were a water equivalent to the mudpot bursts that had captured my attention for nearly 20 years.

From the trail about 100 feet away, the start of an eruption could not be seen, only the water column after it rose close to the crater rim. Years of capturing mudpot explosions enabled me to capture some bursts before the water turned to steam, but I didn't know how nice the bursts were until weeks later. Some of the pictures are shown below.

A full eruption

A large, clear bubble

Bubbles breaking up

Wide flat mass of bubbles

These were nice, however, after seeing the pictures on the next two pages, I became hooked on bubble geysers.

The picture that started my passion for bubble geysers. Looks like a pile of bean bags. The large one on top is so transparent that bushes can be seen behind it, and its base consists of bubbles that are cavitating, turning to spray within the large bubble. The "buffalo" on the left is part of a tree root. The burst is about 7 feet high and 16 feet wide at this point. It is about 2 seconds into a 20 second eruption.

Another exceptional eruption. The bubbles are breaking into a very tall eruption with spray approaching 20 feet. The blue bubble occupies almost all of the larger, right hand portion of the pool. This is one of the largest blue bubbles from this geyser.

Next year, 1997, pictures were taken from trailside. No one else was paying much attention to Botryoidal since Great Fountain Geyser was nearby. My pictures improved after seeing those taken in 1996, but it was still difficult to capture the right moment without being able to see the bubble before it domed up or broke through the surface of the pool.

botryoidal spring - the early years

I visited Yellowstone every year from 1996 through 2000. Most pictures of Botryoidal were taken from the trail, although I had permission to take pictures and gather eruption data at poolside on a number of occasions. That photography is shown in a later chapter.

During these early years, some geyser eruption characteristics were monitored and recorded. Continuous video over long periods was also recorded. Geyser eruption periods were recorded for a large number of eruptions and much was learned about its different modes of operation.

At first, the geyser was remarkably predictable. The time between eruptions starts is called its period, the start-to-start time interval. This was just under two minutes with a range between 110 and 125 seconds, a variation of only 15 seconds. The total duration was only about 15 seconds, but initial water rise was the spectacular portion of the eruption.

The geyser seemed to cycle between a series of bubble eruptions and a series of normal splashing or fountain-type eruptions. The closeup observations revealed that slow water rises enabled large subsurface bubbles and resulted in the spectacular bubble eruptions that make Botryoidal so memorable.

The next year, the eruption interval had lengthened considerably - by nearly 50%. The duration was also lengthened by about 25%. Burst sizes remained the same, rising to 10 or 15 feet in the largest bursts.

Influenced by many of the GOSA members, including those acknowledged earlier, I recorded eruption characteristics, principally periods and durations for a ***very large*** number of eruptions (800). This also helped prepare me to photograph each subsequent eruption.

Over the years, this has proved to be informative and useful. A chart of eruption intervals shows that the periods began to level off after a year. The vertical lines through the data points show the range of intervals associated with the average data points The extreme regularity in 1996 is clear from its very short vertical line.

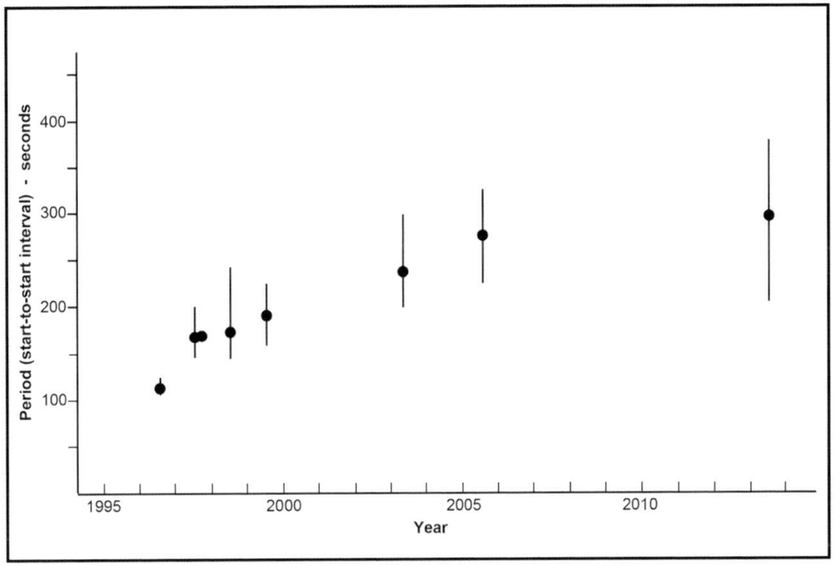

A table of data in support of this chart is shown below. It covers all observations from 1996 through 2013. Once eruption durations stopped increasing and leveled off in 1997, little duration data was collected.

Date	# Periods	Average	Std Dev	Min Period	Max Period	Duration		
						Ave	Min	Max
		sec	sec	sec	sec	sec	sec	sec
Aug 96	35	114.4	3.1	110	125	14.6	10.5	17.1
Jul 97	50	167.1	12.6	147	196	19.1	14	23
Oct 97	106	170.4	12.6					
Aug 98	86	172	16.5	145	244			
Aug 99	120	186.7	12.8	163	225	19.2	13	25
Jun 03	45	239.6	25	196	293			
Sep 05	114	276.2	19.6	223	326			
Aug 13	283	294.4	27.1	206	378			

The data table shows that in 1996, the total data collection time (number of periods times average period time) was an hour, increasing to 23 hours for the extensive observations of 2013.

These first years of photography at Botryoidal were a learning experience. The pool surface wasn't visible until the water had risen and begun to dome up, signalling the start of the eruption. Video turned out to be the only practical way to observe an eruption sequence from the start. In effect, it was only possible to obtain one photo for each eruption.

On the other hand, the stop-action digital photography available in 2004 revealed hidden detail in geyser eruption patterns that I hadn't seen before, except in mudpot bursts.

Some pictures were taken a few feet up the hill alongside the trail. This afforded a view of the pool. The water would rise at least 6 inches in the half-minute before an eruption, therefore making it easier to photo the eruption start.

Pictures from trailside follow on the next few pages.

An unusual three part burst that is overflowing the crater. Water is beginning to flow down the junctions between each section of the burst.

A complex burst with fine structure all over. The section on the left looks like a stela, the inscribed stone pillars made by the Egyptians, the Mayans and other civilizations of old. It is also reflected in the overflow pond.

A large massive burst rises uniformly across the pool. It is just starting to cavitate all over the top, turning into spray that will change picturesque textured bubbles into normal geyser spray.

The combination of early morning sun and gentle breezes drifting the steam away can lead to beautiful eruptions. In addition, this one shows the initial breakout from the bubble plus a nice reflection as well.

Another eruption at the instant of bubble breakout. Several unusual features are visible: concentric furrows all around the bubble surface and a dome of white, steamy froth inside, leaving a transparent region above this dome.

Velvet Spring is located in the Shoshone Geyser Basin at the edge of Shoshone Lake in the south backcountry. It erupts some years, occasionally producing irregular bubbles from the front of its two eruption vents.

other bubble geysers

Only two of the world's geysers reliably produce blue bubble eruptions. Bubbles have been observed at several dozen other Yellowstone geysers and underwater "popping" sounds are commonly heard from some major geysers, primarily Great Fountain, Fountain and Morning Geysers in the Lower Geyser Basin. Photographs of the pure blue bubble phase of their eruptions are almost non-existent.

Cell phone video is beginning to produce blue bubble pictures, particularly of Great Fountain, where their wide angle lenses are more practical. Most of my bubble images of other geysers are in my mind, not in my camera.

Some very nice pictures of the late stages of a bubble eruption have been published, most notably Bruce Jensen's spectacular photo of spherical spray spreading out from the top of a huge blue bubble from Great Fountain Geyser. This was the cover photo in the GOSA Geyser Gazer Sput issue of April 2009.

Pat Snyder also wrote a nice article on bubble geysers in the February 2014 Sput containing pictures of Morning Geyser bubbles, Great Fountain and a nice closeup of the bubble breakout from Depression Geyser near Beehive Geyser on Geyser Hill in the Upper Geyser Basin. There is also an eruption sequence from Strokkur Geyser in Iceland.

Geysers with infrequent eruptions far larger than normal can have massive blue bubbles. A typical story: Rocco Paperiello recounts being near Cauliflower Geyser in the Upper Geyser Basin when the water suddenly domed up many feet above its elevated rim. This was followed by a huge burst. I have heard similar stories about Deep Blue Geyser in the Kaleidoscope Group.

Perhaps a future version of this book can include many images and stories of other bubble eruptions including other contributors.

Other than fleeting hints of bubbles from Fountain, Great Fountain and a misty, early Morning Geyser eruption, the only other geysers in which I've seen bubbles were Velvet Spring at Shoshone, Solitary Geyser and Spa Geyser in the Upper Geyser Basin. Spa's bubbles unfortunately occurred after a marathon eruption of nearby Grotto Geyser in twilight conditions.

Solitary Geyser is located near the top of the trail to the Observation Point overlooking the Old Faithful portion of the Upper Geyser Basin. It normally boils up to as much as 6 feet every few minutes, however one season, I was fortunate to see a number of 35 foot superbursts from poolside. These sprayed water beyond the pool in all directions. A typical small burst is shown below. The start of a massive spectacular superburst follows on the next page. I had never seen anything like this before - and right before my eyes as well. A smaller, more typical burst is shown on Page 37.

A textured bubble from Velvet Spring in Shoshone Geyser Basin. It is nicely framed by spray all around its back.

Another burst from Velvet Spring, showing eruption from the rear vent as well. Bubbles are only produced in the front vent. A nice geyserite rim lines the runoff channel, even more picturesque than the "grapes" at Botryoidal.

closeup detail

In 1996, Botryoidal would alternate a series of nice bubble eruptions with a series of spray or fountain geyser-type eruptions. Experiencing an hour without seeing many bubbles was frustrating, however its good times more than made up for those bad sequences. I guess the good times were starting to spoil me.

The interval between eruptions continued to lengthen after the 1998 season. On the other hand, the eruption durations, heights and other characteristics remained the same. Also, it seemed like the percentage of eruptions without visible bubbles was decreasing.

In 1997, I applied to the Park Service for permission to observe and tape eruptions from poolside. My purpose was to identify eruption characteristics that would correlate with these two modes of operation. They provided me with an off-trail research sign to use so that the occasional passers-by would not come visit or go back to report my "transgressions".

Fortunately, there were broad areas in front of the geyser that did not have geyserite shelves, so it was safe to observe from multiple vantage points.

This chapter presents pictures from this and a later time when permission was also obtained from the Park Service to observe from close range.

The view from poolside was magical. Water would start rising more than a half-minute before an eruption. Gurgling would occur as the water passed alcoves in the crater wall. Bubbles could be seen entering the pool from an opening about 10 feet below the surface. As they rose toward the surface the pool would visibly rise.

A significant correlation was observed between speed and amount of rise and quality of the subsequent eruption. Rapid rise and low water levels meant weak or spray eruptions. On the other hand slower water rise to a higher level usually resulted in large, beautiful eruptions.

Years later, the same effect would be observed at Strokkur Geyser in Iceland.

Eventually, a possible explanation came to light for the changes in Botryoidal's period and duration. Its sudden increase in activity was probably due to a precipitous increase in heat inflow. This caused rapid water heating and quick eruptions. Bubbles would make their way to the surface before much water could be forced upwards.

As the heat influx lowered with time, its period lengthened, allowing more time for the water to rise to the high level associated with large bubble eruptions.

On the other hand, once enough heating occurred to form bubbles sufficiently large to reach the surface, the eruption would proceed independently of time needed to heat the water. Therefore eruption durations should level off, as was observed.

The pictures on the preceding page and above show complex bursts that abruptly rise from calm water without disturbing it at all. This shows the rapid rise of the bursts. However, one view shows an outward wave disturbance that must have resulted from a much slower rise.

A strong lateral burst from one side shoots way beyond the edge of the pool. The result is a burst in the shape of a turkey. This bird was featured on a Thanksgiving card I sent out that year in place of my normal Christmas card mudpot or geyser burst.

Shade and mist are usually avoided in geyser and mudpot photography. In this case, the early morning mist hugging the pool produces a soft image that is a nice, pleasing contrast to the sharp sunlit images usually obtained.

Backlighting highlights the waves and breakout spray in this burst from a low-water pool.
More closeup images are shown on the following pages. In them, the water looks more like engraved crystal sculptures.

View from the hill showing Botryoidal, its crater and runoff channel flowing into superhot White Creek.
From this vantage point, the rise of bubbles proceeding an eruption is visible, especially when using binoculars, as is its underwater bubble.

view from the hill

After years of photography from the trail closest to Botryoidal, I chose to start taking pictures from a nearby hill to the northeast. This was farther from the geyser, but was a secluded vantage point and also outside the geyser basin.

It required use of high power lenses, binoculars and tripods for steady viewing and photography. On the other hand, it afforded a view down into the crater so that early bubble rise could be observed.

When I was on the trail, few people ever ventured up from the nearby road. Now I could work in seclusion and the few people passing on the trail below weren't even aware of my presence. It was also much easier to take video, photograph and record eruption data simultaneously. Vertical panoramas on this page plus a two page spread on the following pages show the magnificent view from my lair.

One could also see the upper portions of Great Fountain, all of White Dome and the steam cloud from Fountain Geyser at the same time. Animals would pass by as well. One day, a coyote crossed the creek carrying what looked like a marmot back to the kids for dinner.

A panorama view of White Creek and Botryoidal Spring, looking southeast from the nearby hill. White Creek is flowing from left to right toward a bridge on the one-way loop road that goes to Great Fountain Geyser, less than 200 yards to the right of this vantage point.

The creek is primarily fed by thermal features upstream and is too hot to immerse your hand in it. It is about 5 feet wide.

One year, there was a small, unnamed geyser at the left edge of the picture.

The panorama continues on the next page. Botryoidal can be seen in eruption. Just off the panorama to the right is a group of small geysers including A-0, A-1, A-2 and some unnamed ones. One unnamed one close to Botryoidal is at most several feet high, but has an interesting interchange of function among its vents and pools within a shallow crater. These geysers are more visible from the road. Only the late stages of eruptions from Botryoidal are visible from the road, not its initial bubbles.

Wide angle and closeup views from the hill. An extensive large bubble is still underwater. This could be Santa and his sleigh full of toys. Note how calm the water is adjacent to the rising bubble. More examples are shown on the following pages.
Notable features are: the underwater bubble masses on Page 58; long complex bubble pattern on Page 60; precise, furrowed bubble dome on Page 61; complex underwater shape on Page 62; long bubble dome throwing up a large water backdrop on Page 63 and complex overall shapes on Pp 64 & 65.

Some pictures reveal hidden objects or subjects of special interest. The next 4 pages present a sampling of such images. Here, a giant face stares out, complete with large eyes, round nose, fat cheeks and hair.

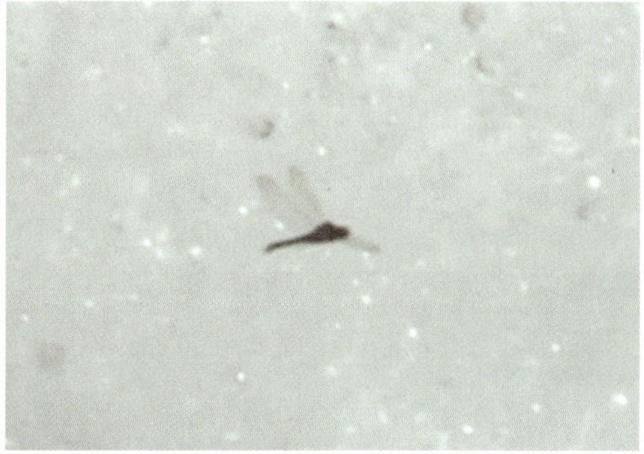

This picture was taken from several hundred feet away. Much to my surprise, a dragonfly was caught flying across the burst. With its huge, multifaceted eyes, it must have had a very confusing view of the eruption.

Two different bubble geysers with identical eruption patterns. Solitary Geyser on the left and Botryoidal above. Each has a round center with lobes on each side.

Botryoidal has a vent and crater size which enable bubbles to spread out and multiply, creating more variation in eruption pattern than any geyser I have ever observed or known. Therefore it stands to reason, that at some point it could simulate many bursts from other geysers erupting from a similar pool.

A fish with goldfish-like tail, chubby lips and cheeks. The tail is produced by a strong lateral lobe of the burst, much like that in the “turkey” burst on Page 44.

This burst has gone on long enough that a wave of water is being pushed outward ahead of the bubble. Also, the clear bubbles on the top right are beginning to collapse. It looks to be a low, wide eruption.

Another unusual burst pushing a wave of water ahead of it. It also has an interesting surface topography with different sized bubbles breaking up all over its surface. An earlier set of bubbles has just turned into a mostly-transparent region with the next set of breaking bubbles visible within. Therefore, its overall appearance is very 3-dimensional.

Unusual flat vertical surfaces all around with a nice textured surface pattern and cavitation just beginning across the top. It has already risen enough to form a very tall burst. A crease in its center suggests that two halves of the burst have merged during its rise.

On a cloudy day, bubble geyser photography shuts down. Just when the eruption started, a patch of sunlight passed over the pool, highlighting this nicely-shaped start of an eruption. One small area is beginning to cavitate (burst into spray) while the rest is still expanding.

The hazy sunlight highlights the striking furrows and patterns across the top of this burst. Its pushing the mound of water higher than usual, overtopping most of the surrounding lip of the pool. Like the previous photo, this is a closeup from the nearby hill.

A pretty eruption. It looks like the bubble-patterned surface is reflected off the quiet pool in front, however its actually a similar bubble beneath the first one.

A two-part bubble is rising toward the surface. What makes Botryoidal Spring the best and most diverse bubble geyser is that this rise is so gentle that bubbles can spread apart and gently mold the eruption into such a wide variety of patterns. These are two separate, but similar eruptions, shown at slightly different times during the eruption sequences.

The timing of when each bubble breaks the surface makes a big difference in the subsequent burst shape. You can imagine what it might look like based on the burst sequences shown in this book.

eruption sequences

Prior to the advent of practical digital photography (about 2004), the most practical way to build a time history of a geyser or mudpot burst was to take video at high shutter speeds and look at it one frame at a time.

The only way I could obtain an idea of the bubble burst progression was to look at a set of photos, one per eruption, then build of a set of pictures from similar burst shapes. This was much easier to do with Strokkur Geyser in Iceland, since its symmetrical bursts do not vary much in form,as will be illustrated in the next chapter.

Picture sequences like the one that began this book are presented in this chapter. The progression from underwater bubbles to blue bubbles, textured surfaces and spray can be quite spectacular.

Note also that prior to seeing bubbles as they near the surface, the water surface lightens due to water rise plus sunlight reflecting off the underwater bubble surfaces. I use this effect to pinpoint the start time for shooting each sequence.

Sequences are introduced by a one or two-image burst plus text so that the longer sequences can be presented as full-page images without text on each page.

Eruption thumbnail sequences are shown on Page 107 for comparison with Strokkur Geyser sequences in the next chapter.

Pp 78-81: Evolution of a rapid rise to a large burst.

The bubble dome above turns into a face, possibly with spectacles, in the second image. Compare it with the face on Page 66.

One of the best sequences. The first four images clearly show the increased reflection on the water surface due to both the rise in surface level and the increased reflection from the underwater bubble. Images 4-8 to follow show a complex, broad bubble, giant splashes and a broad, 2-peaked blue bubble with central furrow for water flow. Look closely at the next to last image. It has faces hidden all over its surface. Good luck hunting.

It looks like an explosion has occurred in the bubble, producing the jets that are rising in all directions, like a fireworks burst.

The following 6 image sequence shows a rare event. It starts with a rather ordinary burst that is quite symmetrical. Then in the last two images, an internal blue bubble can be seen rising through the spray.
This was the first time I had ever witnessed such an event. Later, I captured another one, but this was the nicest.

This pair shows textured surfaces merging into two smooth sections that may merge into a single bubble with a crease or channel at their merging surface.

The first two of the 4 image sequence on the following pages shows spaced bubbles that eventually merge. The eruption starts at a low water level

A bubble perched atop a large bubble mass has yet to breach the pool surface that it has pushed up above the rim of the pool. In the second image, the small bubble has expanded and begun to cavitate into spray. The large mass below is forming a complex surface that has reached the rim of the pool, however it still has a thin film of water over its entire surface.

The following pages show a misty 3 picture sequence in which the middle image looks like a diving submarine. The third image (Page 103) has all sorts of interesting surface textures over the different regions of its surface.

A typical burst is shown here. Its large bubble is quite symmetrical and this uniformity continues as the bubble rises to where cavitation indicates that it has started to break out at its top

On the other hand, the 2 picture sequence starting on the next page spreads rapidly to rebound off all sides of the pool. Three domes then rise vertically to entirely fill the eruption end of the pool.

Botryoidal eruption sequences are shown for comparison with Strokkur sequences shown on Pp 132-133.

Note the large variety in bubble shapes for Botryoidal due to the smaller and slower eruptions, plus the bubbles get to wander around the pool as they rise.

In contrast, Strokkur bursts are so constrained and so much more powerful, that its blue bubbles do not vary much at all.

This closeup view of Strokkur Geyser in Iceland shows a number of significant differences from Botryoidal. First, this obvious vertically-directed burst is far more powerful. Also, visitors have ready access all around the geyser, held back only by a moveable rope barrier close enough to provide a warm shower if the wind is right.

Iceland's crown jewel

Strokkur Geyser is the most prolific producer of blue bubble eruptions in the World. It is located at the south edge of the Haukadalur area of south-central Iceland in the Geysir Thermal Field. The Great Geysir, namesake for which all geysers are named, is located on a low mound adjacent to Strokkur.

The thermal area is very ancient. First reference to it was made in the year 1294. Reference to Geysir specifically dates back to 1647. Strokkur has been known to erupt for 200 years, beginning with an earthquake in 1789. In 1800, it was reported to erupt in a thin column to 200 feet (60 m) continuously for hours at a time, dwarfing the Great Geysir.

An earthquake in 1896 ended its eruptions which by then had lowered to half as high as earlier reported. A reactivation briefly produced small eruptions in 1907, however it soon became dormant.

In 1963, a 130 foot (40 m) hole was drilled down the center of its crater. It successfully reactivated Strokkur to as much as 100 foot heights, making it the major active geyser in Iceland and the second most popular tourist attraction after the Blue Lagoon, a thermal spa adjacent to a geothermal power plant near Reykjavik. The third most popular is the spectacular Gulfoss waterfall just down the road to the east of Huakadalur.

Just across the road from the thermal area is the Hotel Geysir boasting one of the best buffet brunches I have ever experienced. It takes up both levels of the hotel. Adjacent is an interesting museum.

I visited the area for 9 days in late May, 2005, spending most of the time taking photos and video of Strokkur. The following two pages show a panorama of the area from a hill to the north. A rare small eruption of the Great Geysir occurred when making the panorama, so I artificially added it to the picture, simulating what a dual eruption of these magnificent geysers would look like.

Left half of a panorama from a hill north of Strokkur. A unexpected, but small eruption of the Great Geysir occurred. An hour later, Strokkur was photographed in large eruption. It was added to the panorama (adjusting its shadow) to indicate how a dual eruption would appear.

Right half of the panorama. This view to the south from a nearby hilltop shows how barren the landscape is - its actually quite refreshing to experience.

Note the low trees in the area. They were planted a long time ago and form one of the few "forests" in Iceland. The Hotel Geysir and Museum are the red-roofed buildings beyond the thermal area. Many buses pull up during the day for the buffet and a visit to Srokkur. To my mind, that buffet could be the best one in the World. The geyser and buffet make this the second most popular tourist attraction in Iceland.

Its the end of May and a weekday, but almost no people are in the area. On the weekend there will be a small crowd gathering around the geyser, however as you can see by the pictures, weekdays are a solitary experience.

Most of the pictures are taken looking west toward the hills, south toward the hotel or east toward a snow-capped volcano. The runoff channel goes to the south and southwest so the water and steam from previous eruptions prevent closeup pictures toward the north.

Strokkur proved well up to its reputation. It would have series of eruptions consisting of 1 to about 4 separate bursts separated by a pause of 5 minutes or so. A number of eruptions every hour would reach 70 to 100 feet and 10 to 15 blue bubbles could be seen every hour. Since you can watch from right next to Strokkur, these eruptions seem especially powerful, especially when compared to watching Old Faithful Geyser.

Strokkur means the churn in Icelandic. Before an eruption, the water sloshes around, just as it would in a butter churn.

I didn't take rapid sequences of pictures during the eruptions, like those shown for Botryoidal. On the other hand, since the crater is small, symmetrical and punctured by the borehole, the eruptions are quite repetitive - nothing like the extreme variability seen in Botryoidal.

Looking to the west and the runoff channel, the crater is empty with some water flowing back into its narrow vent.
At the right, an "animal?" at the top. Below: a typical way to enjoy an eruption.

This is the ultimate example of a classic blue bubble. It is helped by the drilling out of Strokkur's vent plus the symmetry of its crater. The next two pages show examples of the beautiful pure bubbles that come from Strokkur.

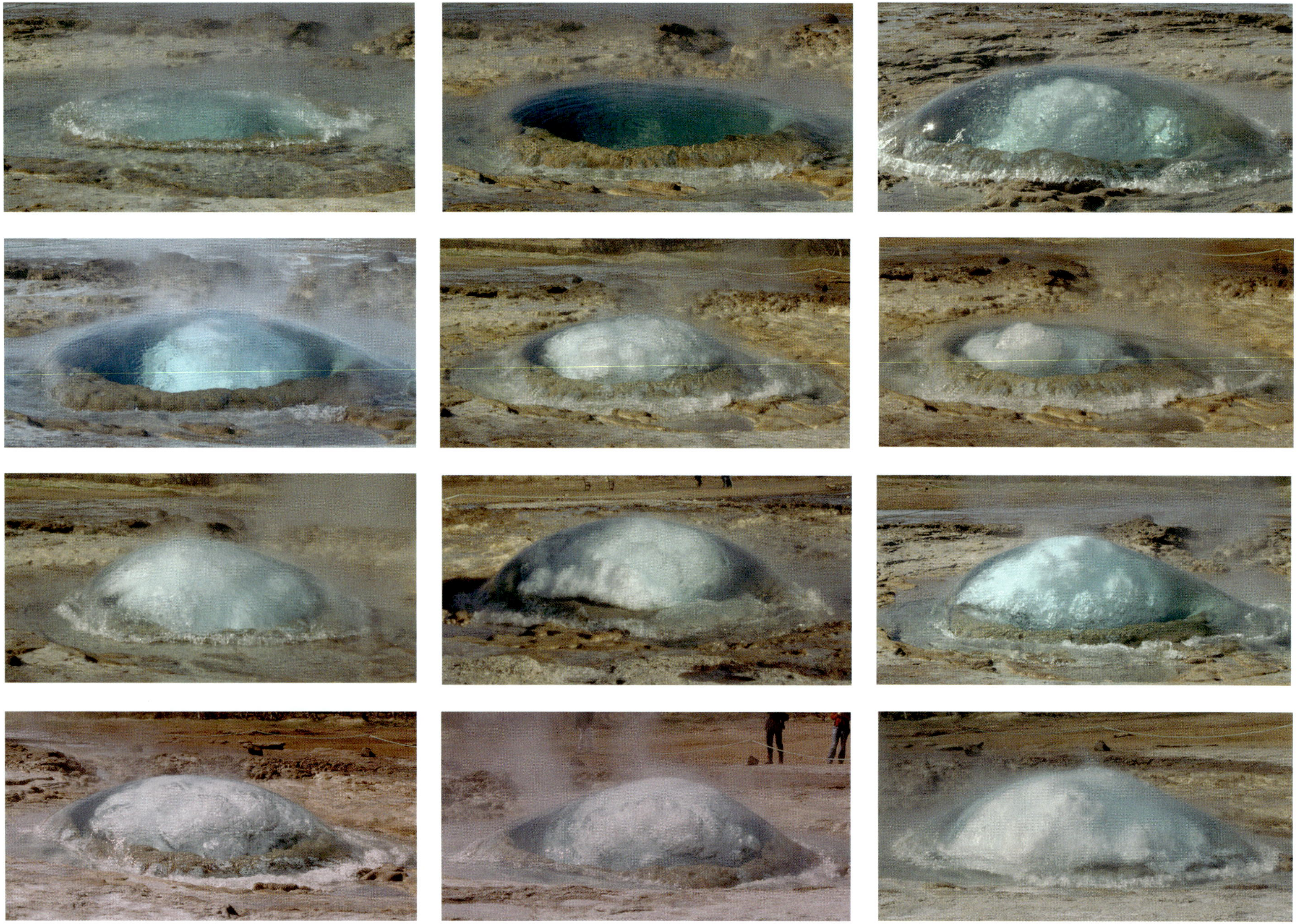

Two more full bursts from Strokkur. The backlit one on the left appears to be more explosively outward. On the other hand, the one on the right is more vertically oriented, except for its low lateral surge to the right. Another interesting feature: there appears to be a shaft of water going straight up to the highest point of the eruption. Could this be an instance of smooth water flow from the central drill hole in its vent center?

A narrow column of spray emerges vertically from a smooth blue bubble.

A forceful eruption from the top of this bubble turns it into a conical shape and creating the appearance of a volcano. The internal steam bubbles are approaching the water surface, producing a strongly-textured surface to the bubble as it breaks apart.

A similar, yet different eruption than on the previous page. This burst has progressed farther and is accelerating upwards.

The bubble is developing a complex structure as it begins to break out. The Hotel Geysir and Museum are the red-roofed buildings.

Internal steam bubbles are at the surface and deforming it. Breakouts from the bubble appear all around its peak.

Steam is nearing the bubble surface all over, breaking out at each end. The strong outflow of water has created splashing all around. Around the rim, areas of cavitation bubbles can be seen through the water flow. They are adhering to the rim surface.

A uniform textured surface, delicate spray breakout and complex outwash wave pattern combine for a beautiful eruption. It has the appearance of a small eruption, because the spray is expanding in all directions at a slow rate.

This looks to be a small eruption that is breaking out all over with gentle puffs of spray. The appearance of this burst can be more readily visualized after seeing results of the special photographic techniques presented in the next subsection of this chapter.

This closeup shows how a bubble turns into a mostly upward- directed burst like the full-size eruptions shown at the beginning of this chapter

I saw many eruptions on this trip and learned to gauge differences in their mode of operation. Three distinct modes were observed: (1) **normal** in which the water did its "churn" effect, a to and fro disturbance followed by rise to overflow the rim soon followed by an eruption; (2) **high water** where strong overflow all around would precede the eruption, often leading to a larger eruption. This effect was also seen in Botryoidal. One day: (3) a **low water** mode was observed. Eruptions started out of sight, more than 5 feet below the crater rim. Nearly all were totally spray eruptions, although the largest ones had partial bubbles of water that could be briefly glimpsed as they zoomed skyward. No photographs were taken, and it was a good excuse to take a break from geyser gazing and take a tour around this beautiful island.

The high speed pictures shown so far show Strokkur as it changes from blue bubble to a rising column of spray.

Late in my trip, I wanted to see what would happen if I slowed down the camera shutter speed and reduced the light reaching the camera through the use of filters. The result came as a pleasant surprise. These pictures showed long steaks of water at locations where the spray is jetting away from the bubble at a high rate of speed.

For example, the high-speed picture on this page and the previous page show certain areas where the bubble appears stationary and other areas where plumes of spray are breaking through the bubble and leaving at a relatively high rate of speed.

Compare them with the slow-speed pictures on the following pages. Although they are different eruptions, there are strong similarities in many of the bubble breakout patterns.

Strokkur seems to be the ideal geyser for these slow shutter speed shots. Its bubble remains relatively static and jets of spray shoot out in various directions. In large bursts, the vertical jetting can be quite rapid, leading to very elongated spray patterns.

Compare this to Page 125 which shows a similar burst with jets shooting out all over its bubble. It was taken using traditional high speed photography.

Symmetric jets shoot out at a high rate of speed, forming a good representation of a fleur-de-lis, the heraldic symbol used in many European coats of arms, most notably the French. It means "flower of the lily"

A low burst spreading out in all directions from multiple breakout locations. See Page 124 for a similar weak breakout all over.

An interesting breakout pattern with strong start at one end and weak at the other. See Pp 120 and 122 for normal photos of similar eruptions.

A nice, complex burst rising more rapidly from its center. Relative speeds of water rise can be determined by the length of the blur in different portions of the bubble.

Two very similar burst patterns: Large bursts rising as energetic eruptions from tall, textured bubbles filling their craters. Each has an offshoot that is starting to form a V-shaped burst.

Rapid sequence photos of Strokkur eruptions were not obtained in 2005 so frames from a video camera were turned into burst sequence for comparison with those shown for Botryoidal on Page 107. The shapes of the blue bubbles do not change much due to a number of factors. First, the crater is small and circular, constraining the water to a circular shape. Second, the 40 m hole drilled in 1963 to reactivate the long-dormant geyser constrains the eruption to a vertical direction. Third, many of the eruptions are far more powerful than Botryoidal, preventing much variation in how the bubble breaks up. Weaker bursts show the most variation. A sequence is shown on the following page. Consequently, the most striking variations in Strokkur burst patterns occur in tall, full eruptions. These pictures change with wind, lighting, visitor antics and background scenery.

Strokkur has a third mode of eruption that was only seen on one weekend of observations. It consists of bursts from a water level far below the crater rim. Unfortunately, no video was obtained. This view of a low-water second burst is identical to weak low-water mode initial bursts. Some powerful bursts from low water also carried a bubble within the spray to a meter or so above the rim. That would make an interesting video to see).

More eruption sequences from Strokkur. The first two look toward a snow-clad volcano to the east. Although blue bubble phases are almost identical, subsequent bursts are quite different. The top one appears quite weak. In the final frame, its bubble is stationary or dropping and the eruption appears to be forming a low, wider one. In the second, the breakout is narrow and rapidly rising with enough force to pull the dome up into a cone shape, signalling a large, tall eruption.

The third sequence has a blue bubble covered with bubbly texture, followed by a pretty breakout from a collapsing bubble, signalling a small burst.

The final one is producing a very wide vertical column which could signal a massive burst, however the blue bubble is not being pulled upwards into a conical shape, so the burst will probably be of intermediate height.

A fat breakout from the geyser bubble culminates in an energetic explosion from its top, spreading out in all directions.

Another nearly identical one more vertically-oriented is shown below.

In January 2015, a tour group visited Strokkur Geyser. Ten eruptions were observed. Only one had multiburst eruptions like many I had observed in 2005. The largest bursts observed were about 70 feet tall and intervals were about 7 or 8 minutes. These numbers are essentially consistent with data collected in 2005.

During this visit, one remarkable event occurred that has not been previously documented within the geyser observation community. Strokkur blew a spectacular "smoke ring"! Large smoke rings are known to be produced by volcanoes, most notably Mt. Etna in Sicily.

This smoke ring slowly rose in the cold, quiet air for a few minutes after the eruption before slowly dissipating.

A fitting way to end this collection of patterns produced by a special group of geysers.

Photo courtesy of Lyn Harper

books by the author

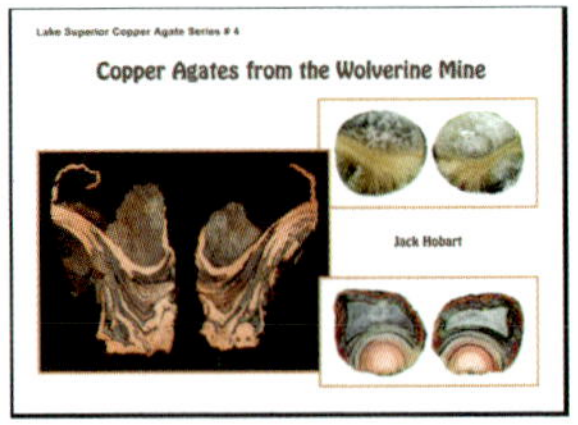

Mexican Opal Canteras
Spectacular Cabs of Fire
2015
softbound
88 pages 133 images

Unusual Geysers,
Hidden Bubble Patterns in Geysers of Yellowstone & Iceland
2016
softbound
136 pages 237 images

Unusual Geysers 2,
Botryoidal Spring Bubble Eruption Sequences

pdf e-book only
224 pages 240 images

Look! What's that Lurking in the Mudpots
"Through the Looking Glass"
2016
softbound
72 pages 117 images

More Things that Lurk in Mudpots
"Witch way did they go?"

softbound
80 pages 156 images

Pocket Mud Geyser,
The Eruption Years 1994-2004

softbound
96 pages 180 images

Skip an Atom Agates
The Most Unusual Lake Superior Agates
2016
softbound
96 pages 276 images

Water Level Agates
Lake Superiors with Surprising Variety

softbound
112 pages 217 images

Lake Superior Copper Agates
Introduction to the World's Most Spectacular Agates
2016
softbound
88 pages 175 images

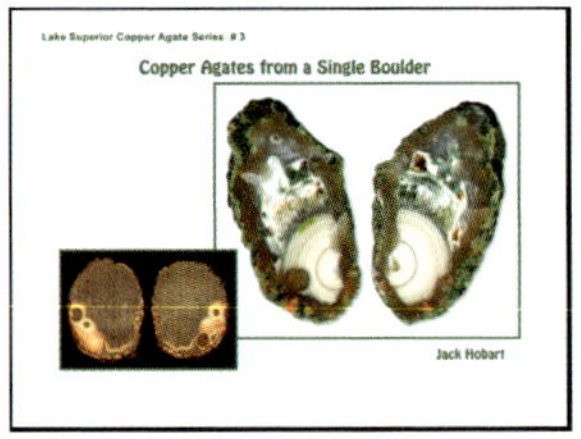

Copper Agates from a Single Boulder

pdf e-book only
88 pages 146 images

Tiny Lake Superior Agates
Treasures Anyone can Find
2016
softbound
64 pages 141 images

Lake Superior Eye Agates
Revealed Like Never Before

softbound
116 pages 179 images